Geometry Colouring Book

Relaxing Colouring with Coloured Outlines and Appendix of Virtue Cards

Deltaspektri

Geometry Colouring Book: Relaxing Colouring with Coloured Outlines and Appendix of Virtue Cards
Musigfi Studio
Copyright © 2015 Päivi Halmekoski

Teachers have permission to make paper copies from the pages of their book for classroom use. They may also project instructions given in the book onto a classroom wall using devices that do not make a permanent electronic copy. Any other copying, including the making of digital copies, is strictly prohibited.

Published by
Deltaspektri
Espoo, Finland

ISBN 978-952-7163-01-6

Contents

Colouring..3
Creating Your Unique Symbols...3
Colouring Images...5
Appendix 1: Building a Tetrahedron..30
    A Ruler and Compass Construction for Drawing the Template.............30
    Folding and Gluing a Tetrahedron..31
Appendix 2: A Sixfold Symmetry Image...32
Appendix 3: Virtues and Virtue Cards...34
    Seeing the Positive in People..34
    Creating Individual Symbols...34

Colouring

Enjoy relaxing and creative colouring moments with these geometric images. Feel free to colour using your favourite technique or to combine different pens or pencils. If you use ink pens, you may want to put a piece of scrap paper under the page you are colouring.

Creating Your Unique Symbols

Some people like to set a topic for the picture they are colouring. Choose one to three words that have a special meaning for you and that you would like to think more about. For example, write "thankfulness, peace, early morning lake" above your image. While you colour, you may begin to connect with the words and build associations with them. You can also colour the words themselves. You can think about their topics in general, or say them silently in your mind while your pen or pencil is moving. If one of the words is a lovely place, you can imagine this place while colouring. Be as slow as you like and take your time – you will create your own unique symbol for thoughts that you value. For example, if you come back to colour a picture about "thankfulness" for five or ten minutes every day, you may be able to think about this topic more often in your daily life. You can find more ideas in the virtue appendix at the end of the book.

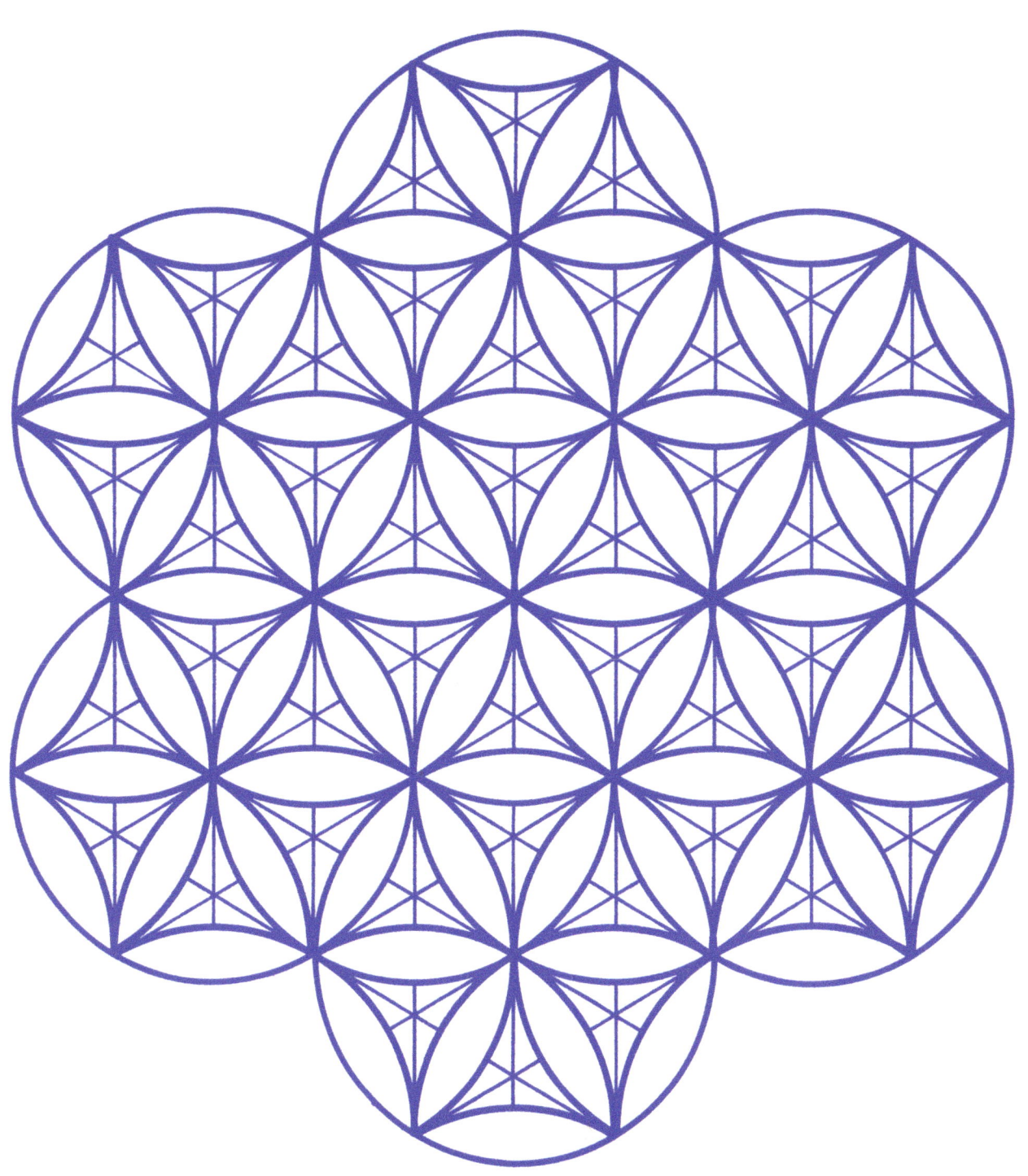

You can fold these into tetrahedrons according to the instructions on page 31.

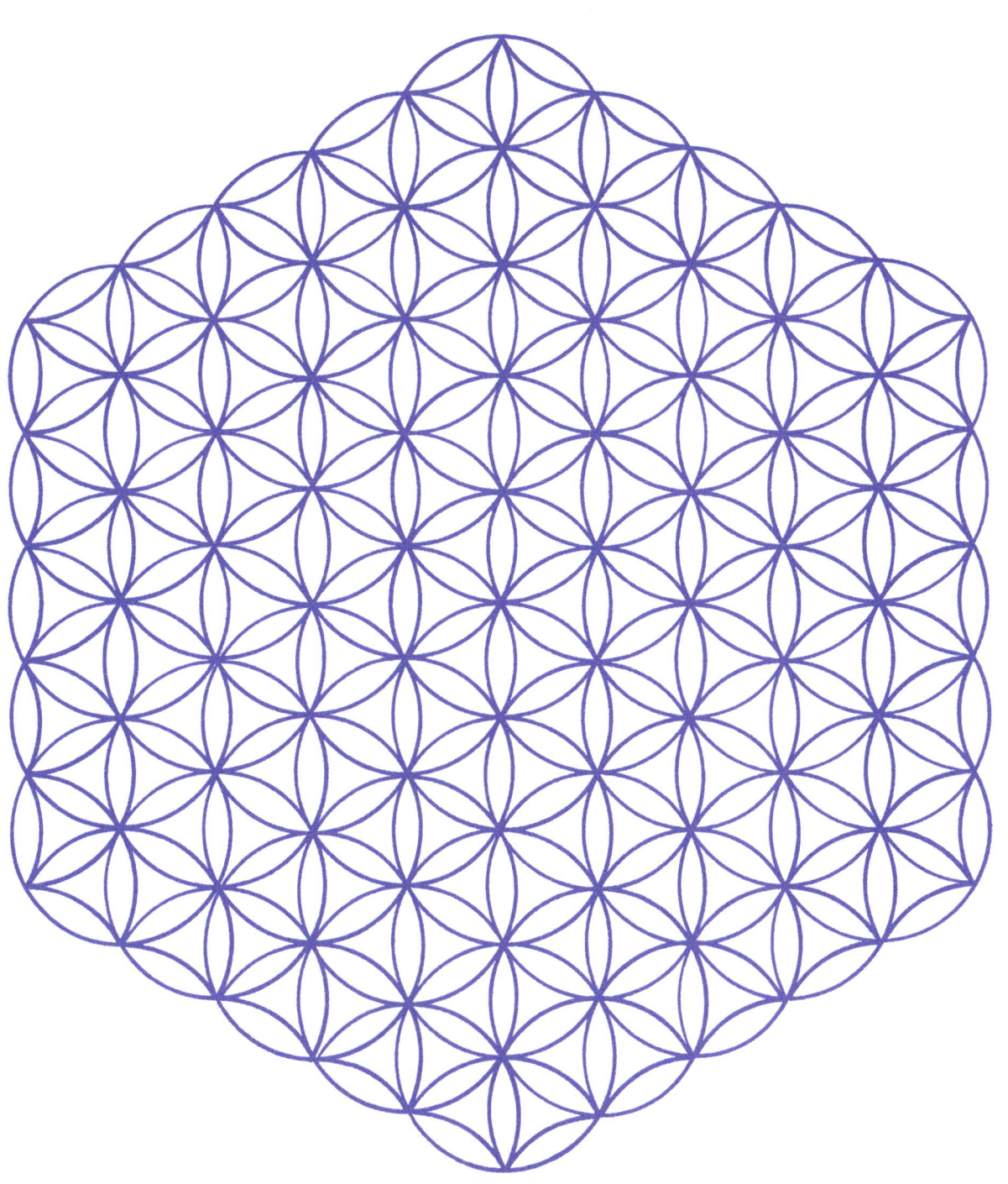

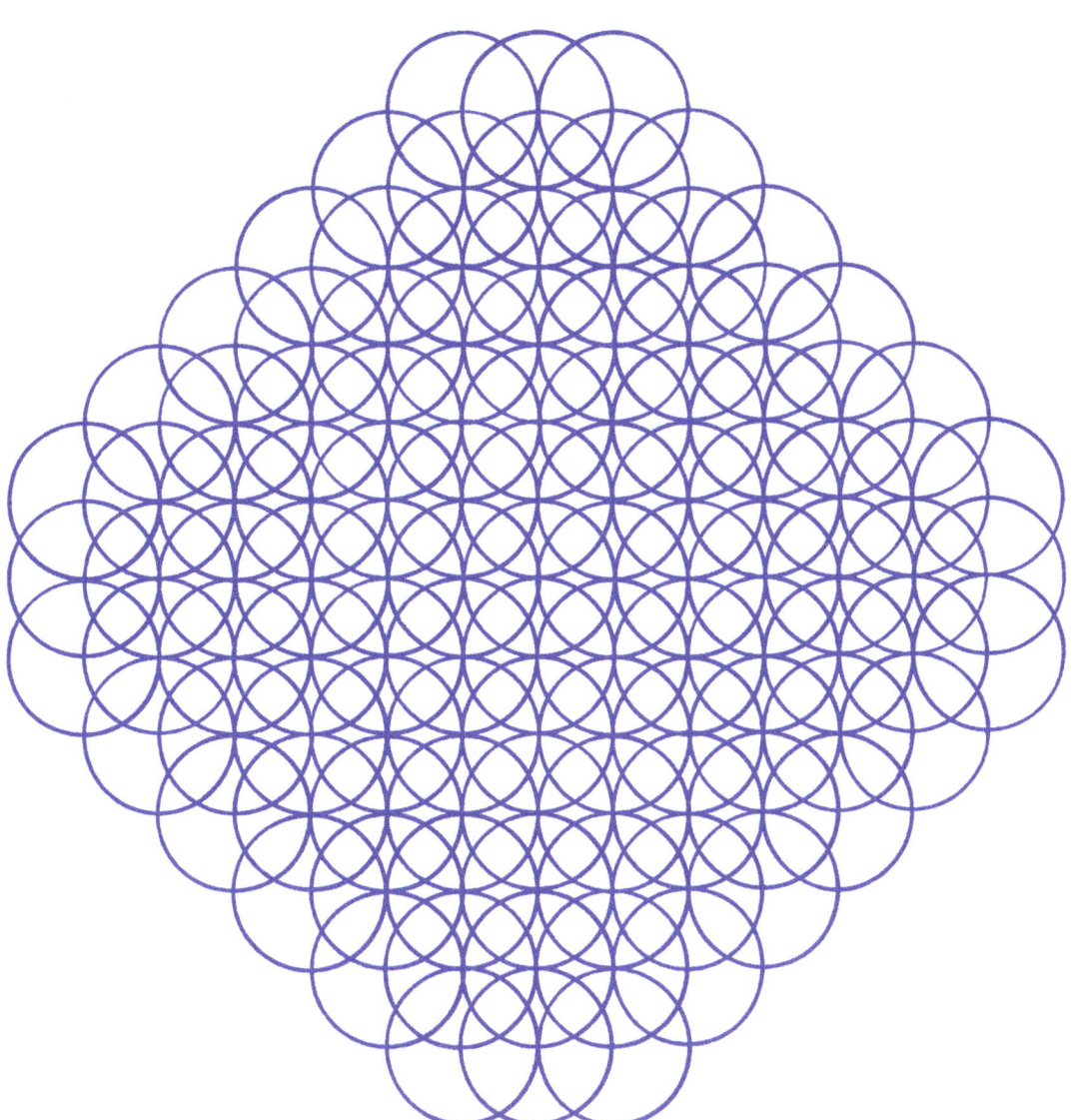

Appendix 1: Building a Tetrahedron

A Ruler and Compass Construction for Drawing the Template

Set the compass to the radius of, for example, 5.12 inches (13.0 cm) for an A4 sheet.

Picture 1. Draw a horizontal line. Draw a point on the line and use it as the origin for a semicircle.

Keep the compass open at the same radius. Move the needle-point to the leftmost point of the diameter of the circle. Draw a short arc that crosses the circle's outline. Repeat on the right side: move the needle-point to the rightmost point of the diameter and draw a short arc that crosses the circle's outline.

Picture 2. Connect the points with a ruler.

Picture 3. Place a ruler over the image along the dashed line in picture 3. Draw a mark in the middle so that it crosses the side of the triangle. Repeat symmetrically by drawing a mark on the opposite side.

Picture 4. Set the radius of the compass to the distance between the origin of the semicircle and one of the marks drawn in the previous step. Place the needle-point of the compass on any of the vertices of the triangles. Draw short arcs that cross the sides of the nearby triangles. Repeat until you have drawn all the marks in the picture.

Picture 5. Connect the points to form smaller triangles.

Picture 6. This is the area that you need for a tetrahedron template. See instructions on the next page for completing the solid.

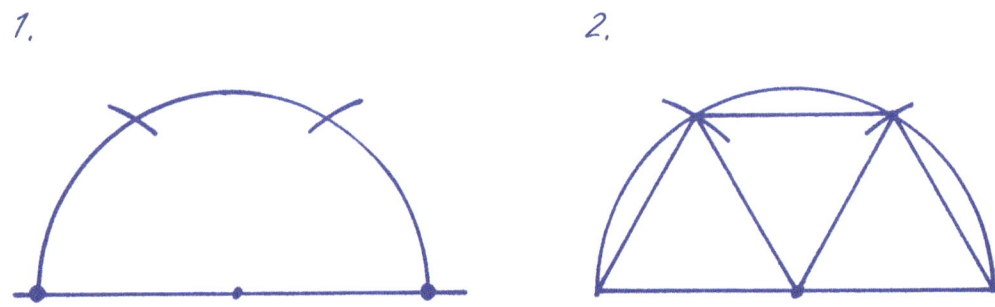

3.

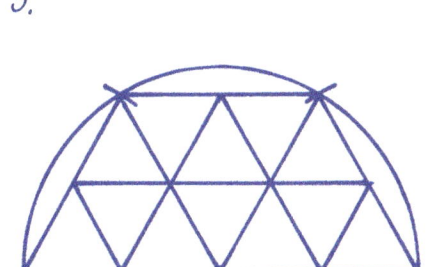

4.

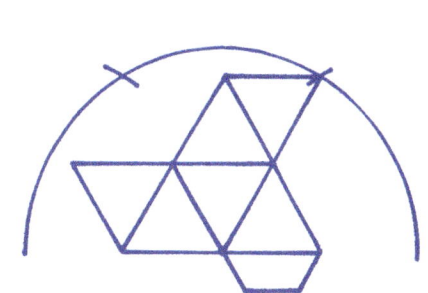

5.

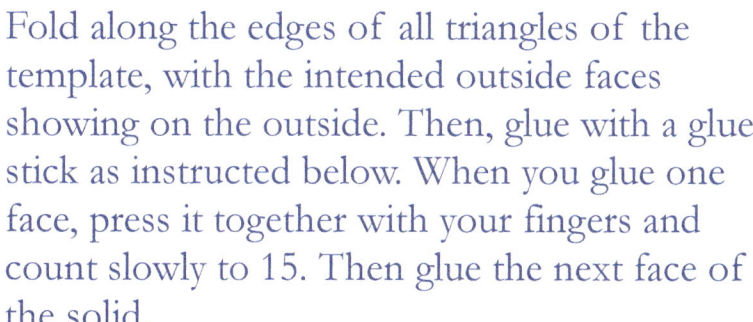

6.

Folding and Gluing a Tetrahedron

Start by cutting along the outer edge of the template. Write the numbers on the picture below on the back of the template.

Fold along the edges of all triangles of the template, with the intended outside faces showing on the outside. Then, glue with a glue stick as instructed below. When you glue one face, press it together with your fingers and count slowly to 15. Then glue the next face of the solid.

Glue quadrilateral 1 under triangle 1. Then glue triangles 2 on top of each other. Finally, glue triangles 3 on top of each other.

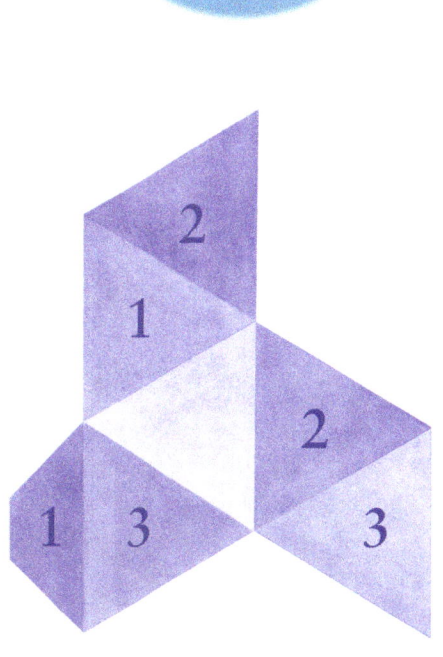

Appendix 2: A Sixfold Symmetry Image

Draw a circle in the middle of the paper, with a radius of about 1.6 inches (4 cm) on an A4 sheet. Keep the compass open at the same radius throughout all of the steps.

Picture 1. Draw a point anywhere on the outline of the circle and place the needle-point of the compass there. Draw a circle. Move the needle-point of the compass to the point where the new circle and the original circle meet. Keep drawing new circles and moving the needle-point to the next intersection point until you have drawn six new circles.

Picture 2. Place the needle-point of the compass at the point where any of the two previously drawn circles meet at the outer edge of the image. Draw an arc that remains wholly inside the image.

Picture 3. Repeat step 2 for the five remaining similar intersection points.

Picture 4. Place the needle-point of the compass on the outermost intersection point of the image. Draw an arc that remains wholly inside the image. Repeat for the five other similar intersection points.

1.

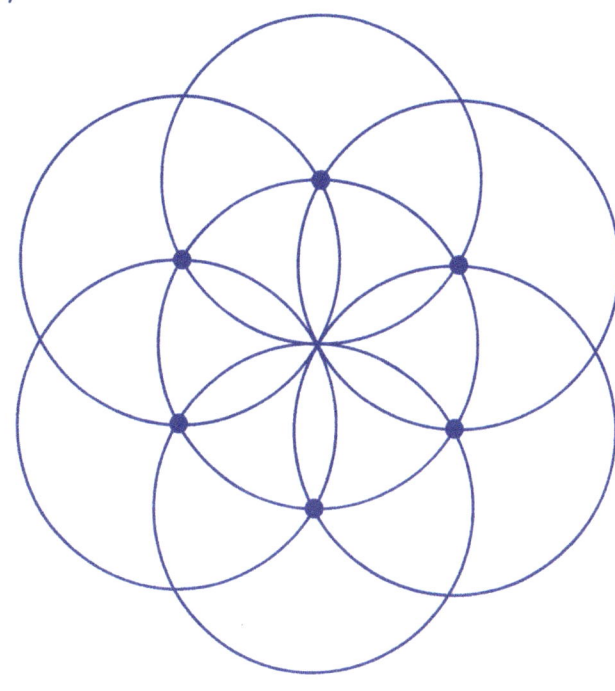

2.

3.

4.

Appendix 3: Virtues and Virtue Cards

Virtues are good qualities in a person's character; qualities that can also be developed. You can use lists of virtues to become aware of your own or other people's good deeds and qualities. By exploring your own good qualities, it may become easier to find the motivation to develop virtues that are perhaps more challenging.

Seeing the Positive in People

Exercise for virtue cards

Choose a friend or group of people. Spread the virtue cards on the table with the text side facing upwards. Bring to mind activities that this person or someone in the group has done and discover acts of virtue in them. Pick up the virtue card in question. Keep recalling virtues and add new virtue cards to the pile of cards now in your hand. Continue for as long as you are able to keep picking cards off the table.

In the second exercise you will practise remembering your acts of virtue. Spread the virtue cards on the table with the text side facing upwards. Bring to mind activities you have done in the past and discover acts of virtue in them. Pick up the virtue card in question. Keep recalling virtues and add the next card to the pile in your hand. Continue for as long as you are able to keep picking up cards.

Creating Individual Symbols

You can choose to set a topic for your colouring picture before you start colouring. Choose one to three words that have a special meaning for you, for example, "thankfulness, peace, early morning lake". Write them either above your image or on a separate piece of paper. You can also cut out the virtue cards on the following pages and use them. Tape the card or cards above your image, or keep them close by on the table where you can see them. Sometimes you may only want to think about them and not have them visible. You can choose to just think about their topics in general, and if one of the words is a lovely place, you can imagine it. Or, you can say the words silently in your mind while you move your pen or pencil. This is especially helpful if you would like to clear your mind. After some practice, it may become possible to think about the words even if you don't see them.

Reliability	Understanding
Kindness	Temperance
Composure	Thankfulness
Diligence	Friendliness
Perseverance	Purposefulness

Courage	Empathy
Moderation	Loyalty
Helpfulness	Justice
Thoughtfulness	Concentration
Orderliness	Honesty

Flexibility	Humility
Creativity	Confidence
Truthfulness	Tact
Patience	Consideration
Enthusiasm	Responsibility

Punctuality	Cleanliness
Peacefulness	Forgiveness
Determination	Respect
Modesty	Fairness
Courtesy	Prudence

Self-discipline	Trustworthiness
Joyfulness	Diplomacy
Cooperation	Caring
Thirst for knowledge	Commitment
Generosity	Calmness

www.ingramcontent.com/pod-product-compliance
Lightning Source LLC
Chambersburg PA
CBHW041659040426

42444CB00021B/3477